NIGHTMARE BUGS!

Giant CENTIPEDES

By Natalie Humphrey

Gareth Stevens PUBLISHING

HOT TOPICS

Please visit our website, www.garethstevens.com. For a free color catalog of all our high-quality books, call toll free 1-800-542-2595 or fax 1-877-542-2596.

Library of Congress Cataloging-in-Publication Data
Names: Humphrey, Natalie, author.
Title: Giant centipedes / Natalie Humphrey.
Description: Buffalo, NY : Gareth Stevens Publishing, [2025] | Series: Nightmare bugs! | Includes index.
Identifiers: LCCN 2023042323 | ISBN 9781538292129 (library binding) | ISBN 9781538292112 (paperback) | ISBN 9781538292136 (ebook)
Subjects: LCSH: Centipedes–Juvenile literature.
Classification: LCC QL449.5 .H86 2025 | DDC 595.6/2–dc23/eng/20231102
LC record available at https://lccn.loc.gov/2023042323

First Edition

Published in 2025 by
Gareth Stevens Publishing
2544 Clinton St
Buffalo, NY 14224

Designer: Andrea Davison-Bartolotta
Editor: Natalie Humphrey

Photo credits: Cover frank60/Shutterstock.com; series art (texture) Video_Stock _ Production/Shutterstock.com; series art (lines) Hananta S/Shutterstock.com; series art (sign) Mr Doomits/Shutterstock.com; series art (trees) Wilqkuku/Shutterstock.com; pp. 5, 17, 29 kooanan007/Shutterstock.com; pp. 7, 9 Krisana Antharith/Shutterstock.com; p. 11 kamnuan/Shutterstock.com; p. 13 Wayan Sumatika/Shutterstock.com; p. 15 Laura Ekleberry/Shutterstock.com; p. 19 SheraleeS/iStockphoto.com; p. 20 Macronatura.es/Shutterstock.com; p. 21 © Matt Berger/iNaturalist.com; p. 23 Vince Adam/Shutterstock.com; p. 25 skifbook/Shutterstock.com; p. 27 popular business/Shutterstock.com; p. 30 SUCHARUT CHOUNYOO/Shutterstock.com.

Printed in the United States of America

CONTENTS

Giant CENTIPEDES!

A centipede bigger than your hand sounds like something from a nightmare. But giant centipedes can be that huge! The Amazonian giant centipede is the biggest centipede in the world. How big can this giant centipede grow? Let's find out!

Terrible Truths

There are many different species, or kinds, of giant centipede. These big bugs can be found in both dry and fairly wet places around the world.

Giant Centipede BODIES

Giant centipedes are huge! The Amazonian giant centipede can grow to 12 inches (30.5 cm) long. Its body is made of **segments**. Each has a pair of legs. A giant centipede often has between 25 to 27 body segments.

Terrible Truths

Giant centipedes breathe through holes in their bodies. These holes can be found behind their legs.

A Giant Centipede's Body

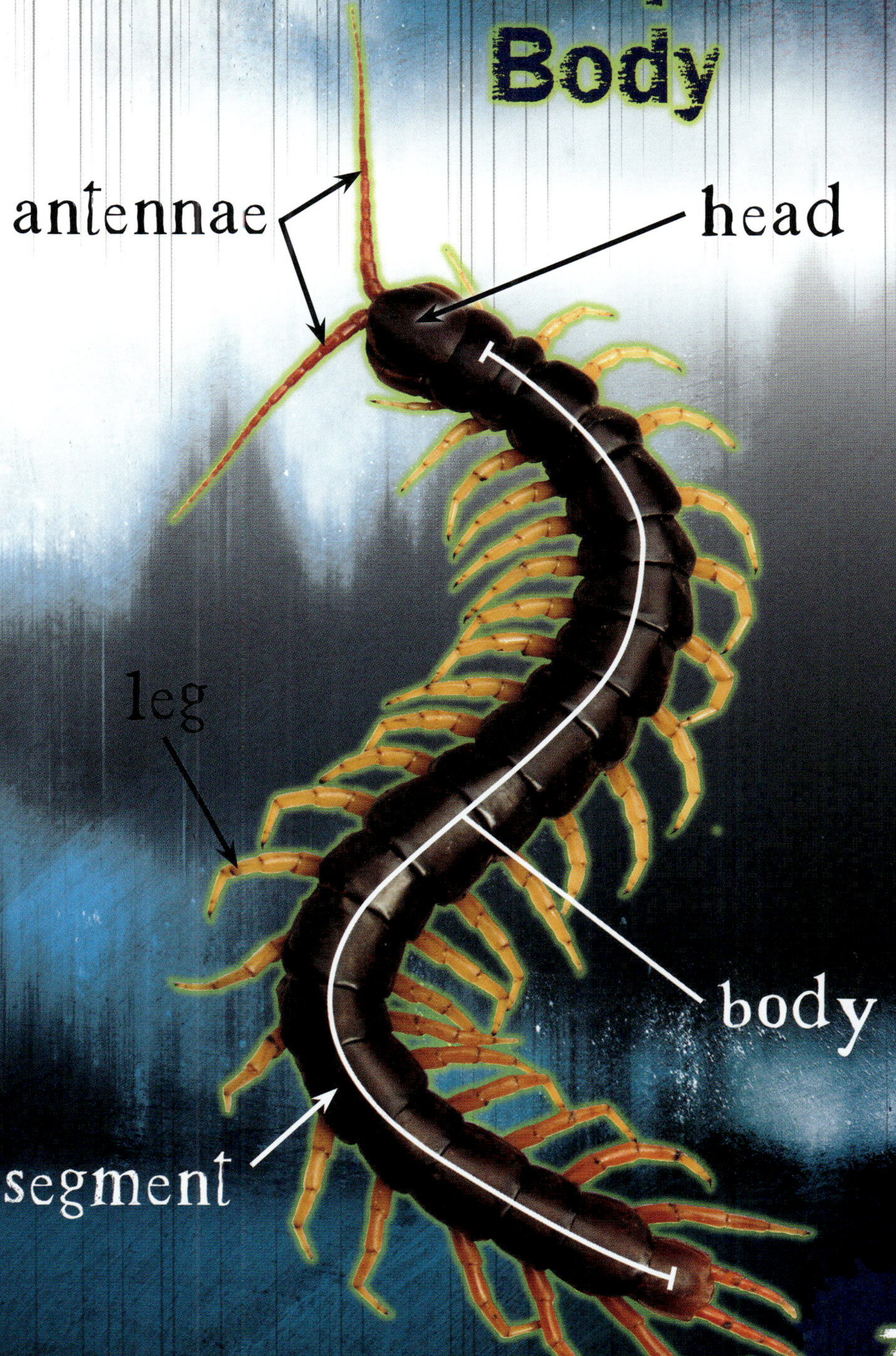

So Many LEGS

Centipedes have a lot of legs! But not all their legs are made for walking. While each of their body segments has two legs on them, only around 21 to 23 segments have legs used for walking. A centipede's other legs are used for sensing danger or hunting.

Terrible Truths

The word "centipede" comes from the Greek words for "100 legs." But many species of giant centipede only have around 46 legs!

Deadly LEGS

A giant centipede's first set of legs is used for hunting. These legs, called forcipules, are more like **fangs** than legs! When a giant centipede is hunting, it uses these legs to grab onto its dinner. The legs **pierce** its food and then the giant centipede uses venom to kill it!

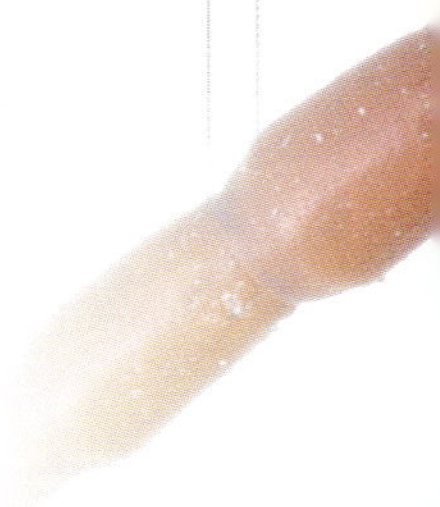

Terrible Truths

Venom is matter an animal's body makes that can harm other animals.

Eating MEAT

Giant centipedes are carnivores. This means that they only eat meat. Giant centipedes usually eat snails, spiders, worms, and any other bugs that get too close to them. But giant centipedes will also eat lizards, birds, mice, and even bats!

Terrible Truths

Giant centipedes use their **antennae** to help them find food. The legs closest to the end of a giant centipede are also used like antennae!

At HOME

The largest centipede in the United States, the giant desert centipede, is found in many southern U.S. states and into Mexico. The Amazonian giant centipede, also called the Peruvian giant yellow-leg centipede, is found in South America and the Caribbean.

Terrible Truths

Giant centipedes can also be found in Australia, Africa, and Asia.

Garden HELPERS

Giant centipedes make their home under dead leaves, rocks, and inside logs. They are great garden helpers. These huge centipedes eat many bugs that could be harmful to plants!

Terrible Truths

Giant centipedes should always be left alone in the wild! Even if they can be helpful, their bites are still painful and can cause sicknesses.

Baby Giant CENTIPEDES

During the summer and fall, a female giant centipede lays around 12 to 60 eggs. She keeps the eggs under her body to keep them safe from predators. She stays with the eggs for around two months before the eggs **hatch**. The baby centipedes look much like adult centipedes.

Terrible Truths

Female giant centipedes will stay with their babies even after they hatch!

When a giant centipede grows, it molts. Molting means it sheds its old, smaller exoskeleton. This makes way for a new exoskeleton to grow! A giant centipede will molt many times before it becomes an adult.

Terrible Truths

An exoskeleton is the hard outer covering of an animal's body.

Run AWAY!

Giant centipedes try to stay away from danger. If a predator or human gets too close to them, they'll try to run away! If they can't get away, the giant centipede will attack. Giant centipedes use their venom to keep themselves safe.

Terrible Truths

Giant centipedes can use their back legs to fight off predators as well. Their back legs have small **spines** and are used to hit nearby predators.

Painful BITES

A giant centipede "bites" using its forelegs, and it can really hurt! While a centipede bite isn't usually deadly to humans, it can feel like a bee sting. Larger centipede bites can be more painful. For some people, the bite can send them to the doctor!

Terrible Truths

Even though giant centipede bites aren't usually deadly, they can be for people **allergic** to centipede bites!

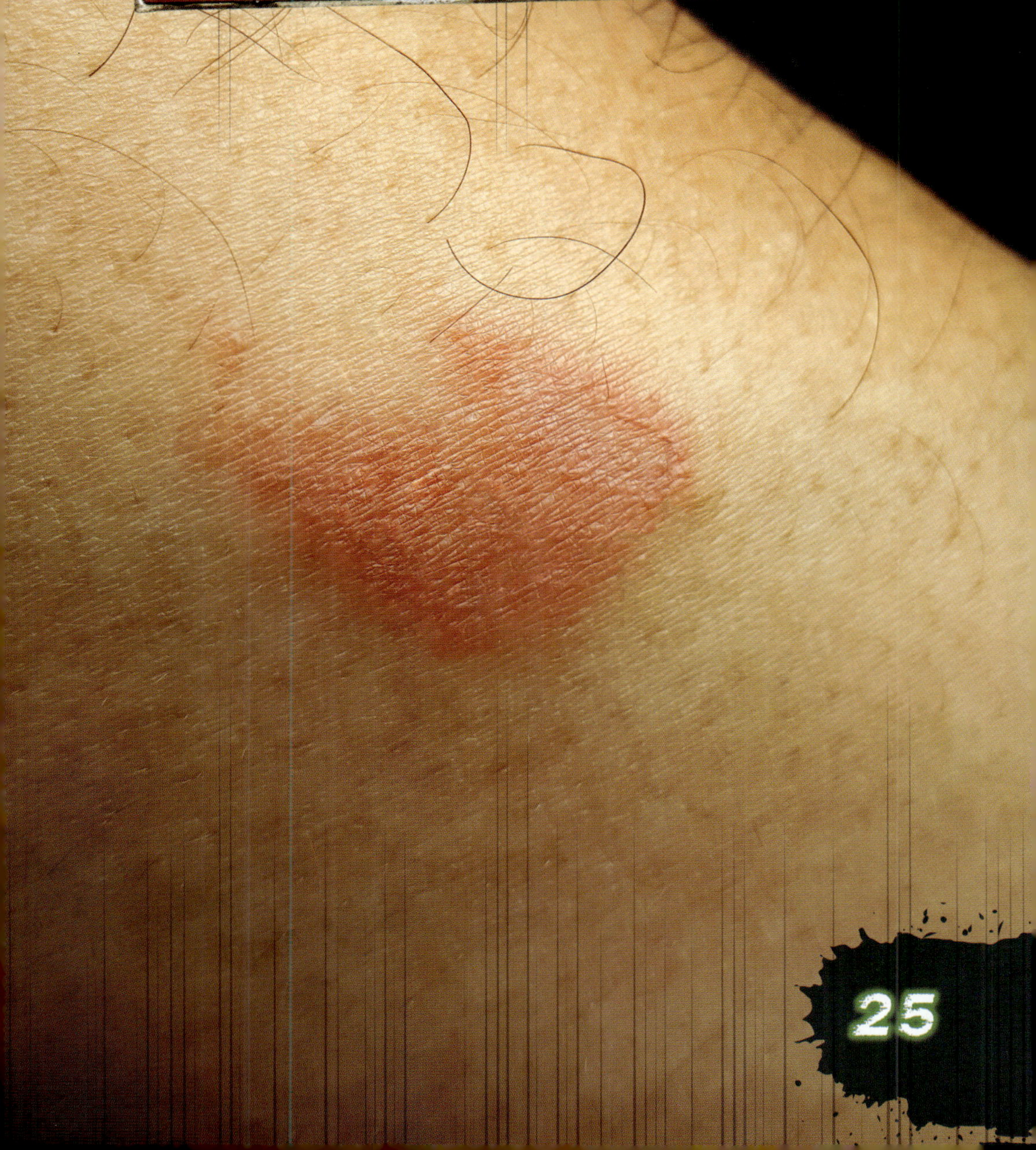

Helping HUMANS

A giant centipede's bite is a painful nightmare, but scientists believe its venom may help humans too! Scientists are studying the way centipede venom works. They believe centipede venom could be used to make **medicine**.

Terrible Truths

Centipede venom has been used in **traditional** medicine for hundreds of years.

Pet Giant CENTIPEDES

Even though giant centipedes can be scary, some people keep them as pets! As pets, they don't need a lot of care. As long as their home is warm, somewhat wet, and they have enough food, giant centipedes can live for a long time.

Terrible Truths

Giant centipedes can live for up to 10 years!

Tiny Terrors

GIANT CENTIPEDES

Size:	up to 12 inches (30.5 cm)
Appearance:	usually dark brown, blue, or orange in color with light-colored legs
Life span:	5 to 10 years
Where Do They Live?	dry or somewhat wet areas with plenty of spots to hide
Diet:	other bugs, lizards, birds, frogs, and bats
Nightmare Fact:	Giant centipedes can eat animals up to 15 times their size

FOR MORE INFORMATION

BOOKS

Hansen, Grace. *Giant Centipedes*. Minneapolis, MN: Abdo Kids Jumbo, 2021.

Hofer, Charles. *Tiny But Deadly Critters*. North Mankato, MN: Capstone Press, 2022.

WEBSITES

Arizona-Sonora Desert Museum: Desert Centipede
www.desertmuseum.org/kids/oz/long-fact-sheets/Desert%20 Centipede.php
Learn more about how the giant desert centipede survives in the wild.

AZ Animals: The 10 Biggest Centipedes in the World!
www.a-z-animals.com/blog/the-10-biggest-centipedes-in-the-world/
Learn more about the different giant centipede species found around the world.

Publisher's note to educators and parents: Our editors have carefully reviewed these websites to ensure that they are suitable for students. Many websites change frequently, however, and we cannot guarantee that a site's future contents will continue to meet our high standards of quality and educational value. Be advised that students should be closely supervised whenever they access the internet.

GLOSSARY

allergic: Having an allergy, which is a sensitivity to a usually harmless thing in surroundings, such as dust or pollen.

antenna: A feeler on the head of some animals. The plural of "antenna" is "antennae."

fang: A long, pointed tooth.

hatch: To break open or come out of.

medicine: A drug taken to make a sick person well.

pierce: To make a hole through.

segment: A part of an insect's body.

spine: One of many stiff, pointed parts growing from an animal.

traditional: Having to do with long-practiced customs.

INDEX